Chinese Cities
Chengdu Impressions

中国城市
成都印象

五洲传播出版社
China Intercontinental Press

A Story of Chengdu
故事
成都传奇

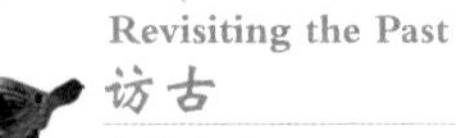

Revisiting the Past
访古
古蜀文明

Cultural Symbols of Shu
蜀地
文化符号

Streets and Alleys
街巷
人间烟火

Art life
艺文
文艺生活

A Story of Chengdu

故事 成都传奇

Kuan and Zhai Lane

Unique Life in Chengdu

宽窄巷子｜最成都的生活样本

This refers to an ancient path built in 1718 during the 57th year of the reign of Qing Emperor Kangxi, which is simple and secluded. Called Kuan (Wide) and Zhai (Narrow) Lane, it is a fine example of the old courtyard culture. The Wide Lane features a slow pace of life full of nostalgia, while the Narrow Lane features a life exquisitely devoted to literature and art. Buildings here each have a history of some 300 years.

这条建于清康熙五十七年（公元1718年）的百年古道，青砖黛瓦、古朴清幽。慢生活的宽巷子充满怀旧感；快节奏的窄巷子精致文艺。这里有老成都的院落文化，又融合着新鲜的生活样式。300年的日升月落、人来人往，这些建筑，斑驳又执着地记录着一代又一代成都人的故事。

Temple of Marquis Wu

Shrine of the Three Kingdoms

武侯祠 | 三国沧桑往事

Zhuge Liang (181-234) lived during the early period of Three Kingdoms period (220-280), wand as Chancellor of Shu Han, one of three states vying for hegemony. The Ancestral Temple of Marquis Wu is located in the southern suburbs of Chengdu, long-time capital of Sichuan Province. It was first built in 223 as a Shrine celebrating the Three Kingdoms period.

锦官城深处，红墙竹影中，隐藏着四川 1800 年波澜壮阔的历史。武侯祠始建于公元 223 年，站在三国遗迹的源头，遥望古老繁华的街区，仿佛穿越时光隧道，昔日刀光剑影如梦，如今竹影摇曳中，一切已是浮云。

Du Fu's Thatched Cottage

Home of the Poet Sage

杜甫草堂 | 诗圣的家园

This thatched cottage lies west of Wanliqiao (bridge). The famed Tang Dynasty poet Du Fu (712-770) lived there for four years. He and his family fled to escape the An Lushan-Shi Siming Rebellion and, arriving in Chengdu in 760, they built this simple cottage. In the four years he lived there, Du Fu composed some 240 poems still cherished today. This part of the world is shaded by woods and bamboo groves, with gentle hills and streams.

万里桥西一草堂，百花潭水即沧浪。草堂四年，是中国被称为诗圣的杜甫（712—770 年）漂泊一生中难得的安定时光。公元 760 年，躲避安史之乱的杜甫携家人来到成都，在浣花溪畔修建草堂。四年间，他写下了 240 余首传诵后世的诗作。这里竹林掩映、诗墨飘香、山水相抱，有着闹市中难得的清幽雅致，也寄托了无数后来者的追思。

Dujiangyan Dam

Man-made Miracle

都江堰｜人定胜天的奇迹

In 256 BCE, Li Bing, the governor of Shu County of Qin State, with his son, started construction of this great project in Sichuan. The stored water was used to irrigate a broad area of farmland, finally turning Sichuan into a "land of abundance" for some 2,200 years. Looking out over Dujiangyan from the Qinyan Tower, one cannot help but marvel at the great wisdom and technical achievements of the ancients.

公元前 256 年，秦国蜀郡太守李冰和他的儿子在四川开始了这项伟大的工程。都江堰不筑堤而化洪涝为桑田，旱涝无常的四川平原，终于成为“天府之国”。蜿蜒的河道，万亩的良田，从此影响了 2200 年的政治格局，造福了千秋万代。秦堰楼上俯瞰都江堰，忍不住惊叹古人的伟大智慧。

Revisiting the Past

访古 古蜀文明

Three-Star Mound

Civilization Sleeping for Thousands of Years

三星堆 | 沉睡数千年的文明

Excavation of Sanxingdui, or the Three-Star Mound, was one of the greatest archaeological discoveries of the 20th Century in China. The bronze masks with eyes protruding forward solves the mystery of Shu king's face featuring the Chan Cong people (whose original image was wild silkworm.) What is the most magical is the sky-scraping God Tree similar to the "passage to heaven" in ancient Chinese legends. Amazing cultural relics unearthed at the site reveal details of the mysterious ancient Shu Kingdom that existed some 4,000 years ago.

这里是 20 世纪最伟大的考古发现之一。眼睛向前突起的青铜面具，解开了蜀王“蚕丛”的面相之谜。最神奇的通天神树，像极了中国古代传说中上天的通道。令人叹为观止的文物，一层层揭开四千年前神秘古蜀国的面纱。

Ruins of Jinsha

Mysterious Ancient Site of Shu State

金沙遗址 | 神秘的古蜀国

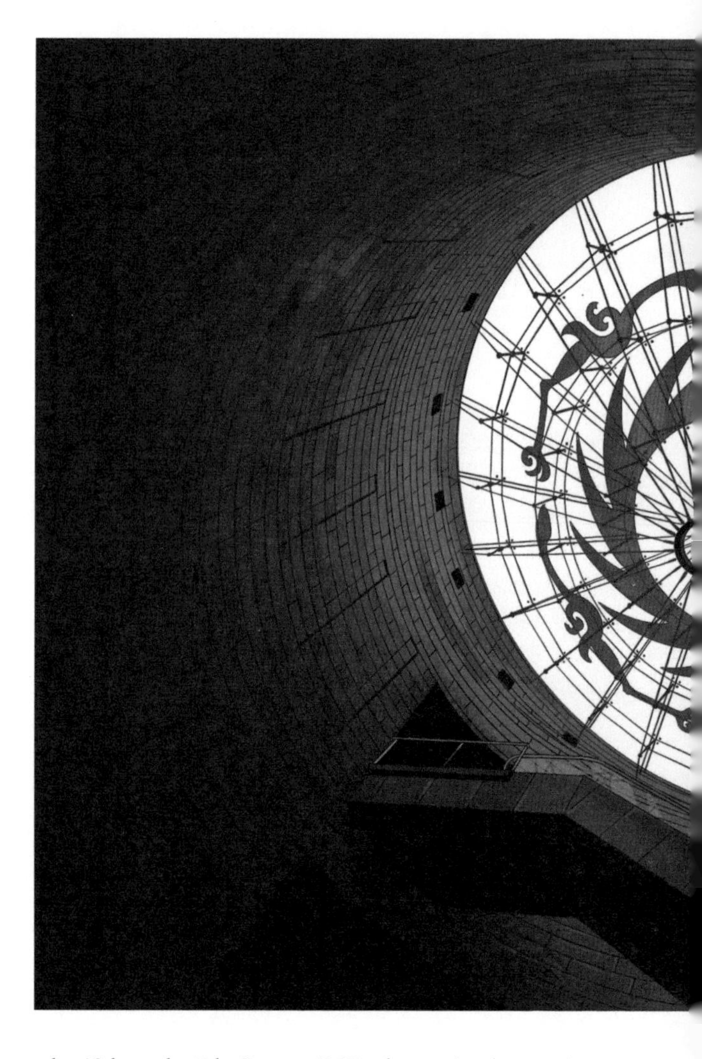

From the 12th to the 7th Century BCE, there existed an ancient civilization centered in the upper reaches of the Yangtze River. This has always been referred to as the State of Shu. People there lived on developed farming and handicrafts. Cities built some 3,000 years ago still shine out from old painted scrolls. These ruins are what remain of Shu civilization.

公元前 12 世纪至公元前 7 世纪，长江上游古代的文明中心一直都是古蜀王国。金沙遗址，见证了古蜀文明最后的辉煌。蜀人农耕有序、手工业精湛，三千多年前的古老都邑，如今仍是华美的画卷。神秘的古蜀文化，就静静地埋藏在这一片遗址中。

Qingcheng Mountain

Overlapping Peaks

青城山｜问道青城山

Qingcheng Mountain forms a series of peaks, with verdant tree coverage. According to legend, this was where Zhang Daoling, the Taoist Heavenly Master, became an immortal. From then on, all the heavenly masters came to worship at the ancestral hall built to commemorate him. Yu Qiuyu, a modern Chinese writer, once wrote a poem extolling the practice of "*worshiping the Dujiangyan and visiting Qingcheng Mountain*". The beautiful scenery of the latter certainly inspires pleasant feelings.

青城山位于群峰环绕起伏、林木葱茏幽翠。传说中，道教天师张道陵在此羽化成仙，从此历代天师都来朝拜祖庭。中国作家余秋雨先生曾在游览时写下“拜水都江堰，问道青城山”意蕴悠远的诗句。青城山，的确是一个山水清幽，令人心旷神怡之地。

Wenshu Temple

Head of the Four Zen Temples

文殊院 | 四大禅林之首

In th 20th year of the reign of Qing Emperor Kangxi, it is said that the Zen Master Ci Du was meditating in the temple ruins and he became aware of being surrounded by lights. People believed he was Buddhist Bodhisattva Manjusri. Over the past 300 years, many eminent monks and people of great virtue have come here to study and visit. Today, a bone relic of Buddha is enshrined in the Chenjing Building, with bones of Master Xuanzang to its left. The temple is still popular.

清康熙二十年，传说慈笃禅师在断壁残垣中结庐坐禅，入定时大放光芒，周边村民惊呼“文殊菩萨”。三百年来，许多高僧大德来这里游学参访。如今，宸经楼内供奉着一粒佛骨舍利，左侧供奉着玄奘法师顶骨舍利。文殊院如今依然香火缭绕，梵音不断。

法鼓

Cultural Symbols of Shu

蜀地 文化符号

Shu Embroidery

Pearl of Shu Creativity

蜀绣 | 蜀中瑰宝

Silk culture is famous legacy of the Shu State, and remains affluent in the Dujiangyan area. Shu embroidery has been passed on for thousands of years. Embroidered lifelike patterns include flowers, birds, insects, fish, and landscapes. Shu embroidery is one of the four famous types of Chinese embroidery.

“蜀”字的虫，就是蚕。都江堰让蜀地沃野千里，适于栽桑养蚕，蜀绣技艺也因此传承千年。花鸟虫鱼，人物山水，目之所及，都能呈现于针尖缎面，呼之欲出，栩栩如生。蜀绣当之无愧的中国四大名绣之一。

Sichuan Opera and Face Changing

Unique Skills of Bashu

川剧与变脸 | 巴蜀绝技

There are two Sichuan Opera masters: Li Longji (685-762), Seventh Emperor of the Tang Dynasty, and Li Cunxu (885-926), an Emperor of the Late Tang Dynasty. The fact that even emperors loved Sichuan opera shows its true charms. Face changing is a unique skill of this particular genre. Peng Denghuai, the noted face changing master, introduced it to Sichuan Opera and raised this unique skill to perfection.

川剧有两位祖师爷，分别是唐朝皇帝李隆基（685—762 年）和后唐皇帝李存勖（885 - 926 年）。能让中国皇帝都如此痴迷，可见川剧的魅力。变脸，更是川剧登峰造极的绝活。变脸大师彭登怀将这一绝活演绎得出神入化，成为川剧变脸第一人。

Giant Panda

Pixiu of Legend

大熊猫 | 传说中的“貔貅”

Eight million years of living fossils have survived the harsh passage of time in nature, only to succumb to the senseless hunting of human. Now, it is forbidden by the law in order to protect these ancient creatures, including pandas loved by all in the world. "Panda diplomacy" has also become a unique diplomatic culture in China.

800万年的活化石，熬过了自然界严酷的优胜劣汰，却险些未熬过人类的疯狂捕猎。如今被立法善待，终于恢复生机。从小到老，无论做什么，都是憨态可掬，一生致力于卖萌。“熊猫外交”也成了中国一种独特的外交文化。

Sichuan Cuisine

Spicy and Delicious

川菜 | 舌尖上的生活

In Chengdu, it seems that every street is flanked by inconspicuous restaurants and snack stands, and the air is full of a strong, spicy fragrance. The food capital of China never hides its love for the good life. Sichuan cuisine, from the main hall of the Qin dynasty bricks and Han dynasty tiles to the people in the alleyways where smoke curls up into the sky, shows the Chengdu people's colorful days.

在成都，似乎每一条街道都铺满不起眼的馆子和小吃摊，空气中漂浮着麻辣的浓烈香味。这座中国的美食之都，从来不掩饰对生活的热爱。川菜，从秦砖汉瓦高堂正厅到袅袅炊烟的巷子人家，绵绵不绝地展示着成都人活色生香的日子。

Streets and Alleys

街巷 人间烟火

Tea Houses in Chengdu

Sipping Tea under Ginkgo Trees

成都茶馆 | 银杏树下 一杯清茶

A gentle breeze and warm sunshine, plus the aromatic flavor of tea - these reflect unique features of life in Chengdu. Old people are often seen playing chess outdoors, and crowds of people are seen sipping tea and chatting harmoniously while children play games around them.

微风浮动、阳光碎影、茶香袅袅，成都坝子上，孩子们奔跑嬉戏，老人们围坐论棋。约上几个朋友，喝喝茶，摆摆龙门阵，是成都人最巴适的时光。

Chunxi Road and Taiguli

Most Prosperous Business Districts

春熙路与太古里 | 成都最繁华的商圈

Chunxi Road is the main shopping area in Chengdu, first built in 1924. The street is nowadays a modern pedestrian zone with shopping malls, stores, restaurants and fast food chains. It is like the tender leaves of an ancient tree, full of vigor and vitality.

众人熙熙，如享太牢，如登春台，故名春熙。春熙路是千年古城里的时尚最前沿，就像一棵古树最鲜嫩的叶尖，处处都是生机勃勃的烟火气。

Jinli Ancient Street

"Riverside Scene at Qingming Festival"in Chengdu

锦里古街 | 成都清明上河图

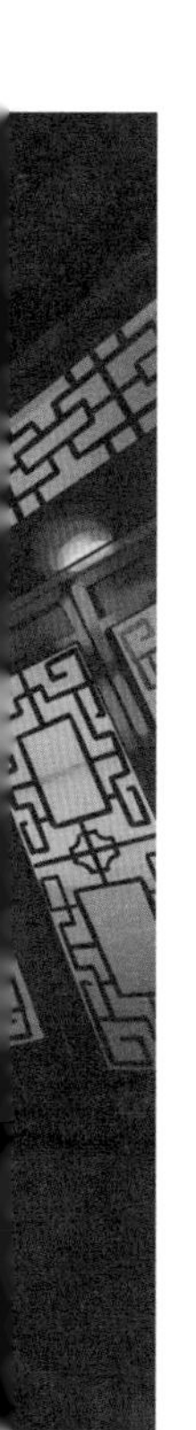

Jinli was a prosperous street in Sichuan that first sprang up more than 2,000 years ago. Everything there, including the architecture and street snacks, show signs of the Three Kingdoms culture. Its night market features a world of Chengdu food specialties.

锦里是两千年前蜀地的繁华街巷。如今这里大到建筑、小到街边小物件，还能依稀感受旧时民风和三国文化。在锦里，传承至今又令人喜闻乐见的，还是街边的小吃。锦里夜市，就是一幅生动的成都美食地图。

Xiaotong Lane and Kuixing Pavilion

Literary Lane Most Favored by the Young

小通巷和奎星楼 | 本土年轻人最爱的文艺小巷

Xiaotong Lane is a famous art Lane in Chengdu. Strolling in the simple and quiet alleyway, one can truly appreciate the leisurely slow life of Chengdu, and experience the unique style of many varied details. Kuixing Pavilion used to be a street of an old mansion, but now it is a place most loved by diners in Chengdu.

小通巷是成都一处难得的文艺小巷。漫步在古朴又清静的小通巷中，感受成都悠闲的慢生活，体会细节处别具匠心的格调。奎星楼曾是旧时公馆一条街，如今更是成都资深老饕们必去的打卡地。

Art life

艺文

文艺生活

Jianshan Bookstore

Hometown of Chengdu

见山书局 | 成都的故里

All questions about Chengdu can be answered in the bookstore, including details of its mountains, rivers, worldly life, history...Located in the Kuan and Zhai Lane, it is often crowded by avid bibliophiles.

这里是成都历史文化专著最富足的一处，你想知道的成都，在这里几乎都能找到答案。“书中能见山见水，见大千世界，也能见古今中外，见世道人心。”来到宽窄巷子，别忘了来这里小坐，书卷在手，开门见山。

U37 Creative Warehouse
Shangri La in Bustling City

U37创意仓库 | 市井深处的“世外桃源”

This is a mini-town of literature and art. There are many paths crisscrossing between the two parallel roads. They lead to secluded places, each offering visitors a surprise. Many are found taking pictures and chatting here.

迷你版的文艺小镇。两条平行的大道间纵横交错着许多小路，曲径通幽处，每个转角都是一个惊喜。清净闲适，一个人散心，两个人闲谈，或者三五相约拍照，这里都是不二之选。

East Suburb Memory
Factory Premises Given New Life

东郊记忆 | 厂房里的奇思妙想

The East Suburb Memory is like the famed 798 art zone in Beijing and Erling No.2 Factory in Chongqing. Abandoned factories and office buildings have been renovated into cultural sites full of life. One can easily make friends with others holding similar ideas in music, fine art, opera and photography.

成都的东郊记忆就如同北京的 798、重庆的鹅岭二厂。废弃的厂房和办公楼在各种奇思妙想的创意之下，变得生机盎然。无论你是热爱音乐、美术，还是戏剧、摄影，在这里都能找到志同道合的小伙伴。

Chongdeli

New Landmark of Fashion and Culture

崇德里 | 成 都 的 回 家 路

In 1925, businessman Wang Chongde built a house here, hence the name Chongdeli. It used to be the Chengdu Branch of the Chinese Anti-Enemy Association and the mint organized by the CPC underground branch. Despite vicissitudes in its history, it has been renovated to become a new landmark in Chengdu.

1925年，商人王崇德在这里建房，遂名“崇德里”。这里曾经是“中华文化界抗敌协会成都分会”，也曾是地下党组织的铸币厂。经历过艰难险阻，也曾破败不堪。特别难得的是，这里并未推倒重建，或是新建做旧。如今的崇德里，修旧如旧，已经成为时尚人文新地标。